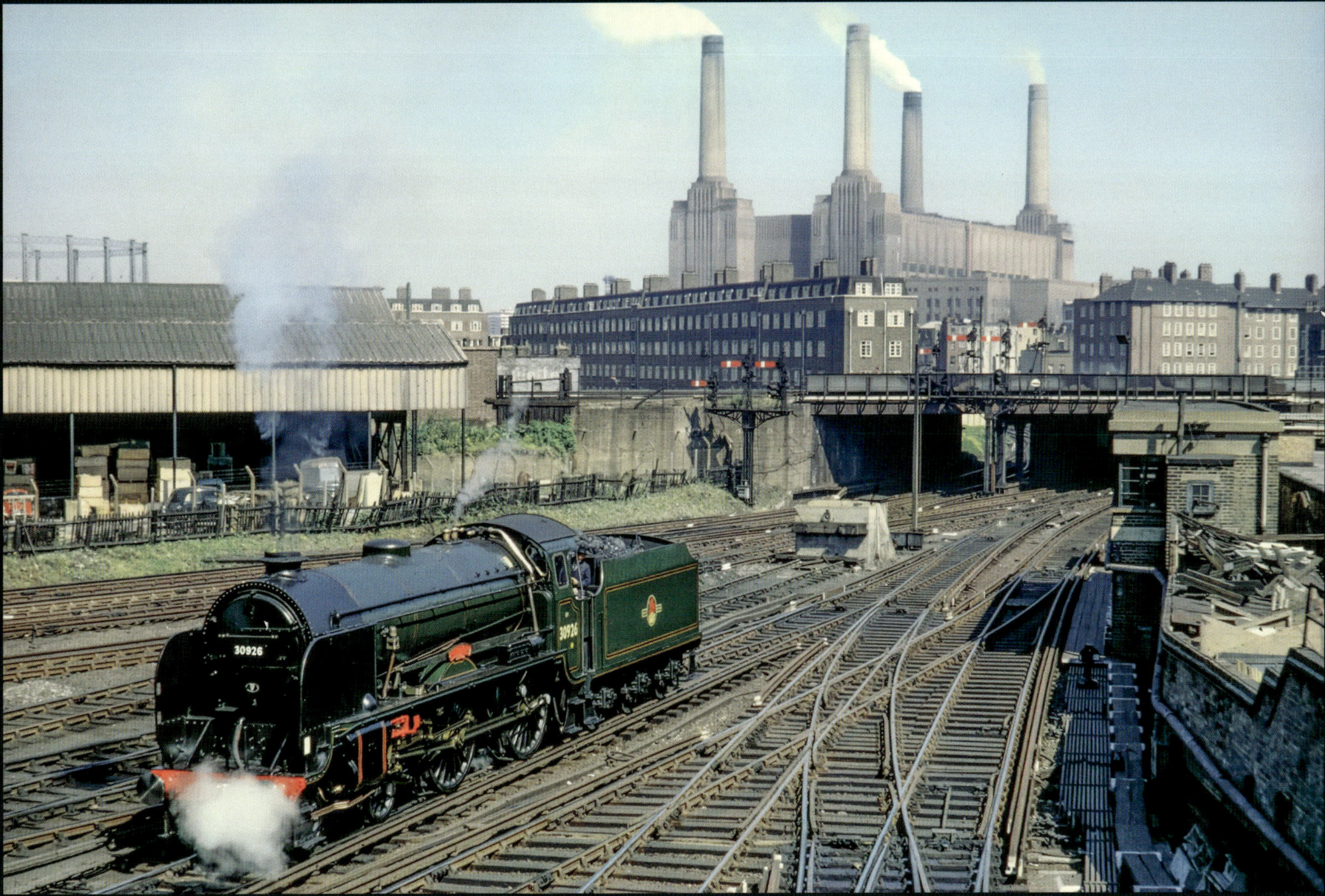

LONDON SOUTH

Images from the Transport Treasury archive

© Images: Transport Treasury or as credited. Design: The Transport Treasury 2025. Text: Jeremy Clarke.

ISBN 978-1-917776-00-4

First published in 2025 by Transport Treasury Publishing Ltd. 16 Highworth Close, High Wycombe, HP13 7PJ
Totem Publishing, an imprint of Transport Treasury Publishing.

The copyright holders hereby give notice that all rights to this work are reserved.
Aside from brief passages for the purpose of review, no part of this work may be reproduced,
copied by electronic or other means, or otherwise stored in any information storage and
retrieval system without written permission from the Publisher. This includes the illustrations
herein which shall remain the copyright of the copyright holder.

www.ttpublishing.co.uk

Printed in Tarxien, Malta by Gutenberg Press Ltd.

Front cover - 'Battle of Britain' class No 34088, *213 Squadron*, gets to grips with the 1 in 61 climb of Grosvenor bank heading the 'Golden Arrow' on 10 September 1960. The engine left Eastleigh in December 1948 for Ramsgate shed. The highlight of its career was, perhaps, in October 1954 when, 'buffed up to the Nines', it worked the Royal Train conveying Emperor Haile Selassie of Abyssinia, visiting as a guest of The Queen, from Portsmouth Harbour to Waterloo via the Mid-Sussex line. *213 Squadron* was rebuilt in April 1960 and withdrawn in March 1967 from Weymouth shed. The Grosvenor Road carriage sheds seen on the right were constructed for the Southern Railway by the Horsley Bridge & Engineering Company of Tipton and completed in 1928. There had been berthing sidings and sheds of various sorts and in several locations on the east side of the Victoria approaches since early LCDR days, though by 1914 nothing was berthed under cover here. Following frequent mockery in the media about the filthy state of the new Southern Railway's carriages, inside and out, this shed was one of several commissioned in the later-1920s. Nine roads are incorporated, each being 750' long: the shed remains in use. *(R C Riley)*

Frontispiece - Maunsell 'Schools' class 4-4-0 No 30926 *Repton* positively gleams as it runs light engine through Stewarts Lane Junction, clearly en route from the shed to power an important working from Victoria station. A trilby-hatted gentleman may be glimpsed on the footplate so we suspect that the royal train might be involved. It's a fine and sunny day in south London but Battersea Power Station seems to be trying hard to spoil it; it is perhaps easy to forget how poor the air quality in the capital was when such coal-burning plants were key to the provision of electricity, located right on its door step. Beyond *Repton* must be one of the sturdiest buffer stops in the land; today, it sits trackless but unmoved between the current running lines and with all that concrete will probably remain so well into the future. *(R C Riley)*

Rear cover - Battersea Power Station had long ceased to contribute to the National Grid when forming the backdrop to class 465 'Networker' No 465915 passing the lines down to Stewarts Lane on 7 July 2011. No fewer than 147 of these units were eventually in service following introduction in December 1992. The contract was complete by April 1995. Three manufacturers were involved and as a result of this, slight differences appear between the various sets. The earliest sub-classes, 465/0 and 465/1, came from BREL and ABB respectively while the third, 465/2, was provided by Metro-Cammell which also built the two-coach 466 class sets. They were supplied in NSE livery and branded 'Kent Link Networker' for use on Eastern Section suburban services. Between 2005 and 2012 all were refurbished, the 34 sets in 465/2 sub class being the first with internal finishing matching the class 375 and including limited first class accommodation to permit them to work outer-suburban traffic alongside that class. They became sub class 465/9. A further general 'refreshment' of all sets began in 2016. *(Andrew Royle)*

INTRODUCTION

South London's railway system is a fascinating tapestry of historical innovation and modern necessity, deeply interwoven with the area's development. To truly understand it we must delve into its origins, its evolution, and the challenges it faces today.

The dawn of the railways saw the Surrey Iron Railway established in the early 19th century and which predates many of the more famous railway lines. While initially horse-drawn, it laid the groundwork for future developments.

It would be the Victorian era that witnessed a surge in railway construction, with the London, Brighton and South Coast Railway and the South Eastern Railway vying to connect the burgeoning suburbs. In turn these lines fuelled South London's growth, transforming villages into bustling residential areas. The intricate network of overground lines that criss-crosses the area remains a testament to this period.

Slightly west but still south of the river, the London & Southampton company had established their terminus at Nine Elms on the south bank of the river. Here was a bustling interchange (the L&S changing its name to the London & South Western Railway even before it reached its original goal of Southampton). Later it would expand towards London proper and with a terminus at Waterloo, this whole extension carried on a series of arches, the underneath of which give rise to much of the culture of South London today.

As traffic increased, helped with the start of electrification in the first decades of the 20th century, so stations like Clapham Junction and London Bridge became pivotal hubs and handled immense passenger volumes. Clapham Junction's complexity remains legendary, making it one of Europe's busiest railway stations.

In recent years London Bridge has gone through extensive modernisation that has greatly increased its capacity.

Originally the South Eastern Railway, London, Brighton & South Coast Railway, and London & South Western Railway operated as separate companies but that would change with the grouping of railways from 1 January 1923, all three merged under the title the 'Southern Railway'.

Whilst electrification was slowly gaining ground, steam traction remained the principal motive power for long distance trains from the various termini. At the same time suburban compartment stock took the stockbrokers and office workers to the metropolis daily returning them in evening; knees touching knees and in the days when smoking was permitted, each individual compartment having its own particular 'fug'.

Steam sheds existed at Nine Elms, Stewarts Lane and Bricklayer's Arms, the origins of these names lost over the decades.

Electric depots were established at places like Wimbledon and Selhurst Park, electric stock having no option but to sit out of use for much of the day and contrary to the wishes of a certain Dr Beeching, who strongly disapproved of limited use vehicles – but then he was never an electric commuter. Working in and out of the various Southern termini were such trains as the 'Golden Arrow', the 'Night Ferry', the 'Brighton Belle' – who can recall the furore over the kippers and Sir Lawrence Olivier in the 1970s? - and the 'Bournemouth Belle'. Most of these would slowly give way to electric traction or fade into obscurity.

Today the modern railway south of the Thames retains much of the flavour of the past but of necessity has had to change to suit the times. Unfortunately there is simply not the opportunity to build further new lines and in consequence it is the responsibility of Network Rail together with the respective operators to maximise the available resources. One example has been the bringing into use of the former Eurostar platforms at Waterloo for general traffic and establishing an efficient Overground service to complement the Underground. Ironically too it is the Underground that is expanding more than the surface lines.

Changing times have also seen former steam and goods depots given over to new uses. Where once there was ash, coal and dirt at Nine Elms, it is now the site of Covent Garden Market. London's railways have and continue to evolve notwithstanding overcrowding being a perennial issue.

Right: On 25 August 1862 the London, Chatham & Dover Railway began working into Victoria station and by the end of the year it had reached the City from Herne Hill to a temporary terminus at Ludgate Hill. The South Eastern had also taken steps to get to the West End, supporting the ostensibly independent London Bridge & Charing Cross Railway, which opened between these two points on 11th January 1864.
At the same time the SER, having noted its rival's move towards the City, obtained authority for a branch from the LB&CCR, bridging the river to its own City terminus at Cannon Street. This line and station opened on 1 September 1866, the SER having gained ownership of its progeny in 1863. Like much of the City the station suffered quite badly during the blitz, among other things losing almost all the glass in the roof and suffering many fires from incendiary attacks. Proposed development saw the roof dismantled in 1958 and the hotel frontage go, two years later. All that remains of the original now are the two brick towers facing the river and parts of the flank walls. Here, on 5 April 1957, not long before moves for redevelopment began, 34076 *41 Squadron* prepares to leave at 4.15pm with the 'Man of Kent' on its 80-minute non-stop run to the first stop at Folkestone. It would then work its way up the coast to Sandwich with a 6.2pm arrival time, and continue as ECS to Ramsgate for servicing. The Up train had left Sandwich at 10.14am with a 1.30pm arrival at Charing Cross. The Kent Coast Electrification saw the service cease. *(R C Riley)*

The 'Golden Arrow' leaves Victoria for Folkestone Harbour behind BR 'Standard' Britannia class Pacific No 70004, *William Shakespeare*, one of two of the class allocated to Stewarts Lane depot at that time for haulage of the prestige expresses between the Capital and the Channel ports. This followed its exhibition at the Festival of Britain on the South Bank near Waterloo in 1951. Despite being the 'odd ones out' in terms of Southern Region loco stock, it was not until 1958 that no 70004 was transferred away, to Trafford Park in Manchester, part of the Longsight District. The engine was withdrawn in 1967. Of equal interest is the gangwayed bogie luggage van behind the tender. The first of these appeared in March 1930 as a result of the frames of many ex-LSWR coaches remaining perfectly usable after the bodies had been taken for conversion into electric stock. The first fifty were joined by forty more from November that year though in these, there were variations in construction. The first twenty-four were, like the originals, 51'3" in length, but on 49'0" frames from 8-compartment 'compos' lengthened by 2'0". The remaining sixteen came out at 53'3" on 51'0" brake and 'compo' frames similarly lengthened by 2'0". The final thirty appeared with the same mix of body length on similarly lengthened frames. Withdrawals began from the late-1950s and all but a few taken for service use had gone by May 1962. *(Flint and Harbart)*

A Maunsell 'N' class 2-6-0 passes through Beckenham Junction with a train for Ramsgate. The number '15' on the lower disc may be the Duty number, though in the list of Engine Workings a BR standard class 4MT 4-6-0 is noted as taking it. The position of the sun relative to the duty's timings, also suggests the train is running slightly earlier than the duty's noted 12.35pm departure from Victoria. Taken in conjunction with the no 174 on the smokebox door it is probable this is a Summer Saturday 'extra', perhaps departing at 11.46 and running as a 'relief' to the regular 11.35. The 'N' dates from the prototype built at Ashford in 1917 and, as a very successful product, was later chosen to be produced at Woolwich to keep the Arsenal employed after cessation of hostilities in the First World War. The Southern purchased fifty sets of parts together with boilers from outside manufacturers. Many were posted to Exmouth Junction from where they found great favour in the West Country. *(Ken Coursey)*

Slades Green was, at one time, a major engine shed for supplying motive power for local South Eastern Railway services. Proposals for its construction were first aired in 1898 as an alternative to expanding Bricklayers Arms where the goods yard was also requiring additional space. It came into full use in 1901 with the title Whitehall, Erith, though the familiar Slades Green was soon adopted, the final 's' disappearing later. It was designed to accommodate up to one hundred and ten engines, principally various tank classes, but as electrification spread the need for this number and for the depot itself declined. Its reincarnation as an EMU depot was decided upon in 1924, soon after Grouping. A large repair shed was constructed close to the engine shed which was finally deserted by its steam stock on commissioning of local electrification in 1926, though it remains in use for EMU inspection. It is pictured here on 8 November 1958 with 4SUB and 4EPB units present. *(R C Riley)*

Kent Electrification was the cause of the procurement of the Birmingham Railway Carriage & Wagon Company 1550bhp Type 3 locomotives for the Southern Region. The 1160bhp Type 2 that became class 26 formed the basis, the additional power being obtained by removing the train heating boiler and installing an 8-cylinder engine in place of the Type 2's 6-cylinder one. The thinking was that the Southern would only use the class for Summer season passenger traffic when heating would not be necessary. Ninety eight units were built by BRCW with Crompton Parkinson electrical equipment that gave them the name 'Cromptons'. Introduced in four 'lots' the numbers were D6500-97, the first engine being delivered on 30 January 1960. An order placed a year earlier included the twelve units built to the narrower loading gauge required by the line between Tonbridge and St Leonards, a group that acquired the pseudonym 'Slim Jims'. One of these, No 33202 and one of the original batch, No 33025, stand with an anonymous member at Hither Green shed on 11th March 1978. These extremely versatile machines were both vacuum and air brake fitted, could work with electric stock, provide electric heating and work in push/pull mode. For those reasons a number of them remain in main line service. *(Bernard Mills)*

Left Top: Resting at Victoria, the 1600hp/600hp Electro-diesel No 73138 is one of two providing power to this train that checks track alignment and level and locates defects for later maintenance, unless those are deemed serious enough to demand immediate attention. The experimental infra-red camera whose efficacy is being tested may be seen between the cab windows and is believed sensitive enough to carry out such checks at up to 90mph. Some similar trains on the routes heading north and west out of London use a converted HST set, its 125mph capability being necessary to carry out the tests without unduly delaying the normal service trains. *(Andrew Royle)*

Left Bottom: Five coach Class 376 'Electrostar' unit No 376018 runs into New Cross on 29 September 2011 on a Down working from Cannon Street, passing an Overground train for Highbury and Islington in the bay platform. The 376s were ordered by Connex to supersede the 465 and 466 classes which were then cascaded to permit withdrawal of the slam door class 423 stock. The first came into service in May 2004, a development of the Class 375 with wider doors and fewer seats to enable quicker movement on and off the unit and maximise space within it. To this end there are no toilet facilities, which may be seen as a retrograde step but which also limit the trains to journeys of not more than one hour duration. All thirty-six sets, numbered 376001-36 were in service by the beginning of 2005 and are based at Slade Green depot but may also travel to Ramsgate for routine maintenance. The operator, Southeastern, is planning a 'mid-life refreshment' programme for all thirty-six units beginning in Spring 2025. *(Andrew Royle)*

Right: British Rail class 378 'Capitalstar' unit No 378256 arrives at Norwood Junction working a London Overground Highbury & Islington-West Croydon service on a bright, crisp 27 January 2012. These units were ordered from Bombardier in 2006 and first appeared in July 2009, originally as three- or four-car sets. The fifty-seven on order were all in service by 2011, the additional vehicles to bring them to five-car sets being provided in 2015. The incentive to acquire these arose from passenger dissatisfaction with the aging fleet on the Silverlink service over the North London line and on the basis that upgrading the older vehicles would be of less benefit than buying new. The units have wide metro-style sliding doors, wide gangways between coaches and fully-longitudinal seating to give more standing room to reduce overcrowding. *(Andrew Royle)*

The 'South Londoner Railtour', which ran on 20 April 1958, was climbing from Stewarts Lane to Factory Junction when a 4SUB unit forming a Victoria-Orpington via Herne Hill train overtook it. The tour had just set out from Victoria with the prospect in its near 4½ hours on the road of visiting places as far apart as Beckenham Junction, Epsom Downs and Wimbledon, as well as travelling oddments such as the Norwood Spur, New Cross Gate to Old Kent Road Jcn, and the branch to Merton Abbey. 'H' class 0-4-4T No 31521 is known to have provided the power for some, if not all of the tour. Battersea Power Station dominates the scene. *(A E Bennett)*

'West Country' class 4-6-2 No 34016, *Bodmin*, passes through Borough Market Junction with the 7.34am from Margate to Charing Cross on 21 April 1958. The train has travelled around the Kent Coast, calling at all stations to Dover Priory but then served Folkestone Central and other principal centres en route to make an 11.37am arrival at Charing Cross. *Bodmin* was among the earliest of the 'Light Pacifics', built at Brighton Works to Order No 2421 and appearing as No 21C116 in November 1945 with allocation to Exmouth Junction shed. The engine went into Eastleigh Works for rebuilding early in 1958 though it had already been reallocated to Ramsgate, perhaps in anticipation of the move there in its rebuilt state: the additional weight made the 'rebuilds' less welcome in Devon and Cornwall. As the engine left the Works in April 1958, it cannot have been long there when taking this duty. A year later *Bodmin* went to Bricklayers Arms before the final move to Eastleigh shed in May 1961 following completion of the Kent Coast Electrification Scheme. Withdrawn in June 1964, 34016 was rescued from Woodhams at Barry and privately purchased eight years later, After restoration, the engine returned to service on the Mid-Hants Railway. *(R C Riley)*

An unidentified 'Bulleid Pacific' leads the 'Golden Arrow' boat train through the semi-rural landscape of outer-London suburbs. From 1952 the train's usual morning departure from Victoria had moved to 2.0pm with a 9.20pm arrival at Gare du Nord. Sailing had also been switched, to Folkestone, though the Up working still set out from Dover. The Arrow did not permit carriage of 3rd class passengers, only the 9.0am service, which berthed at Boulogne, doing so. The lower orders were otherwise obliged to go via Newhaven-Dieppe or Southampton-Le Havre at the risk of mal de mere on the much longer sea crossings. The French were first to apply luxury to the landward part of the journey to London, the 1st class Pullman 'Fleche d'or' coming into service in 1926. The Southern Railway took time to catch up, the English equivalent appearing in 1929, at the same time as a new ship, the 'Canterbury', was commissioned. The company could not have chosen a more inauspicious time to provide such luxury. By 1931 the growth of air travel and the weakening economy virtually forced the Southern to add 2nd and 3rd class accommodation to retain the viability of the train. Wartime saw it withdrawn until resumption on 15 April 1946. Electric haulage took over on completion of the Kent Coast Scheme in 1962, but with the advances in air travel providing a far greater range of Continental destinations at affordable prices the fading glamour of the 'Golden Arrow' - in which toward the end only 1st class had Pullman accommodation - could not compete. The final runs were on 30 September 1972. *(Ken Coursey)*

A fine panoramic 1930s view southwards from Grove Park Station in which the northern end of the Chislehurst tunnels may just be discerned in the distance. A 'V' – Schools'- class engine on the Up Main heads a train of 'narrow' carriages that has most likely come up from Hastings, while a 'C' class 0-6-0 stands in the sidings curving away alongside the branch to Bromley North with an ECS train that it is probably due to haul into Cannon Street for the evening rush hour service. Elsewhere a set of 'Birdcage' coaches is stabled by the Down Main and on the left some further service is provided by two grounded coach bodies: coal waits to be unloaded from wagons in the up side goods yard. The splendid ex-SER bracket signal lasted until the second phase of the Kent Electrification Scheme rendered it redundant from 4 February 1962. Incidentally, the bland present day view gives absolutely no impression of how busy the layout once was. *(Arthur Mace)*

Quadrupling of the line between Bickley Junction and Swanley as part of the Kent Coast Electrification Scheme is well advanced with only weeks to go before the new track comes into use. Maunsell 3-cylinder 'N1' class 2-6-0 No 31876 approaches St Mary Cray Junction with a train for Victoria. The engine was the first of the production 'N1s' completed at Ashford in 1930. The prototype, No 822, left Ashford in March 1923, the fifteenth of an order for 'N' class 2-cylinder 'moguls' but completed as a 3-cylinder engine with a conjugated valve gear invented by Harold Holcroft, who had come from Swindon to Ashford as Assistant CME, and was pushing for its adoption at the time. The five production 'N1s' featured Walschaerts gear from the first and No 822 was later fitted with it to conform. When photographed No 31876 was based at Hither Green shed but had moved on to Stewarts Lane when withdrawn with its five brethren in November 1962. *(R C Riley)*

Parcels traffic at one time brought the railways a steady income until Government permitted a strongly competitive market to be established that naturally favoured the flexibility of road transport. Here, 'D1' class 4-4-0 No 31735 sets out from London Bridge against a background since changed beyond all recognition, with the 12.4pm vans to Tonbridge via Redhill. It is leaving from the earliest part of the station, the site of the London & Greenwich Railway terminus dating from 1836. The engine was built as a 'D' 4-4-0, the last of the ten turned out by Sharp Stewart, in November 1901, and one of the twenty-one rebuilt in 1921 by Beyer, Peacock from the Wainwright 'D' class. At this time it was allocated to Bricklayers Arms but as with many following the Kent Coast electrification of 1959, it found its way west, to be withdrawn from Eastleigh shed in 1961. *(R C Riley)*

With the relatively limited mileage of its main lines, the Southern never ran TPOs with apparatus for exchanging pouches on the move, a service first provided by the Grand Junction Railway in 1838. Neither did it finish its mail vehicles in Post Office Red but in green as here when photographed at London Bridge on a chilly 21 January 1960. The engine, class 'E4' 0-6-2T No 32557, allocated to Bricklayers Arms, is providing steam heat to the train. Among the last of its class to be built, it left Brighton Works bearing the name 'Northlands' in September 1901 for its first shed, New Cross. Inevitably, it moved about as suburban electrification spread, being at Horsham in 1946. It was one of the several members of the class to spend time on empty stock duties between Waterloo and Clapham Junction before withdrawal from Nine Elms in December 1962.
(R C Riley)

Not a hint of high visibility clothing is perhaps the main impression in this view, taken at Hither Green in the aftermath of the derailment of 34084 *253 Squadron* on 20 February 1960 when it had overrun signals with a freight train. Dark boiler suits and overcoats were the order of the day whilst the next move in the recovery of the Battle of Britain Pacific is considered. Where would they be without all those wooden sleepers? Today of course, nobody would be allowed on a site like this without full orange clothing, helmets and all manner of other protective gear. Though steam traction was beginning its final years, this locomotive was still regarded as valuable enough to go through a subsequent visit to main works; should the accident have occurred two years later, the decision would probably have been to scrap it. Notice the billboard advertising Esso petrol with its rather amusing promotional tag line. (R C Riley)

Displaying the No 28 headcode - Hither Green Sidings and Feltham via Richmond - Bulleid 'Q1' class 0-6-0 No 33002 leads the Feltham breakdown train beneath Clapham Junction's 'A' signal box on 4 May 1963. The engine left Brighton Works as 'C2' in May 1942 for Guildford shed where it remained until posting to Feltham about 1960. The signal box itself, 'East' until 1936, was the cause of a major blockage on Monday 10 May 1965 when the northern side at the London end collapsed and fell for about 3½ feet to obstruct the Windsor lines. The cause was determined as corrosion of the steelwork combined with the heavy weight of the steel shielding erected over the cabin roof to protect it from bombs and incendiaries in wartime. That was immediately removed as part of the repair process. Despite a serious fire in June 1986 in which the frame and much wiring was damaged, the box, after the necessary repairs, continued to function until 25 May 1990 when the Wimbledon Signalling Centre took over its responsibilities. The 'Q1' had long gone by that time, being withdrawn in July 1963. *(David Idle)*

Below: Norwood Junction shed with a 'W' class 2-6-4T and 'N' class 2-6-0, No 31826, commanding the centre. The inevitable English Electric diesel shunters are present: these covered up to twenty different duties ranging as far afield as Wimbledon, South Lambeth and Redhill. The Southern's General Manager, Herbert Walker, had an intense dislike of 'light engine' movements and was known, when dissecting the mishap involving one, to be more concerned about the reasons the engine was 'light' in the first place. The new shed at Hither Green had come into use in 1933 to service engines serving the marshalling yards south of the station, thus eliminating the necessary light movements from Bricklayers Arms. For the same reason and built on much the same lines, Norwood shed came into being in 1935 to save light engine running from New Cross Gate and West Croydon. It consisted of a precast concrete and asbestos building with a northlight pattern roof and five dead-end roads. A 65' turntable and water softener were provided at the northern end of the site, with a covered coaling stage reached by a ramp on the west side. The loss of traffic from the mid-1950s and the rise of diesel shunters saw the general allocation of about forty steam engines decline till the shed closed in January 1964. The site is now occupied by the engineers. No 31826 was the first of the fifty class members erected from parts made at the Woolwich Arsenal, entering service in June 1924. Many of its Woolwich contemporaries spent most of their working lives in the West Country at Exmouth Junction and its remoter sub-sheds in Devon and North Cornwall where they were universally popular. 31826 however remained in London, spending many post-war years at Bricklayers Arms until Stewarts Lane took it in until withdrawal in August 1963. *(George W Smith)*

Right: We've become accustomed to seeing images recorded from the top of 'The Shard' but photographer Dick Riley went to obtain this from the top of the previous development on that very same site, Southwark Towers after it was built in 1975. In 2008, this acquired the dubious honour of being one of the tallest buildings to be demolished in the UK, making way for the even taller and more dramatic edifice that we have today. It is rather surprising to see only the one train on the approach to London Bridge station here – perhaps this was taken on a Sunday morning. London Bridge signal box stands out to the right of the station throat; this too had only recently been completed (in 1974) but is no longer in use since the re-modelling of the whole station. *(R C Riley)*

The crew of Bulleid 'West Country' Pacific No 34005, *Barnstaple*, clearly had time for discussion when it was pictured on 12 April 1958 at Cannon Street with a train for Ramsgate via Chislehurst and Chatham. The days of this service in this form were numbered, for electrification of most Kentish lines had been approved as part of the 1955 Modernisation Plan and work on Stage One, which covered the main ex-LCDR routes, had already begun. As from Monday 15 June 1959 this and similar workings would consist of brand new 4CEP and 4BEP stock. *Barnstaple* had been rebuilt at Eastleigh in June 1957 and then posted to Bricklayers Arms shed rather than returning to the more appropriate Exmouth Junction. This anachronism arose mainly because reconstruction added weight that some of the SR's outposts could not carry comfortably. Withdrawal for No 34005 came from Bournemouth shed in October 1966. Incidentally, this engine had been involved in the 1948 Exchange Trials, hauling test trains between St Pancras and Manchester with marked success. *(A E Bennett)*

Stanier '8F' 2-8-0 No 48649 winds its way past Platform 17 at Clapham Junction with a Willesden to Norwood freight. It had arrived here by way of the West London Railway and Latchmere Main Junction, named after Latchmere Road, the A3220, which crosses, north to south, both the South Western and Brighton main lines to/from Waterloo and Victoria. The West London then and for years beforehand had had no scheduled passenger services other than the Kensington Olympia-Clapham Junction shuttle and 'inter-regional' workings on Saturdays at the height of the summer season. However, freight traffic abounded. The '8Fs totalled 842 in number, built over the period 1935 to 1946. Not all were constructed at the LMS company's works, more than a quarter of them being ordered by the Railway Executive Committee between 1943 and 1945 to answer the demand for wartime motive power. These came from railway works at Swindon, Darlington, Doncaster and all three Southern ones. Earlier in the war years North British and Beyer, Peacock had built 208 examples for the War Department. Furthermore, 68 engines were ordered by the LNER of which twenty-five came from Brighton in 1944. Fifty one LMS-built engines were requisitioned in 1941 for service in Persia, though not all went there. Others saw service elsewhere in the Middle East, as well as Italy and Turkey. No 48616 was the first on home soil to be withdrawn, in 1960, one of twenty six in that year. But great blocks of them followed in the next three years and the remaining 150 were all gone by the end of 1968. Fourteen are known to be preserved. *(Larry Fullwood)*

In September 1955, 'W' class 2-6-4T No 31915 rests in the shadow of the viaduct carrying the South London Line over Stewarts Lane. The line opened in stages and was complete by 1 May 1867. Its main claim to fame lies in the LBSCR's efforts to halt and perhaps reverse the sharp decline in passenger numbers as a result of electrification of the local tramways. The eight miles and fifty chains of route between Victoria and London Bridge were provided with overhead catenary at 6700v a/c current from 1 December 1909. The success of the system saw further installations completed and more planned before the onset of the First World War caused a delay that ultimately ended the systems future when the Southern's adoption of the LSWR's third rail 660v system superseded it. It was largely as a result of this widespread electrification in the suburban area that the 'W' class came into being. Not only was the passenger service more frequent than in steam days but the improved acceleration and braking that came with it meant the service as a whole was much faster. Add to that the curvature and gradients of routes across London much used by freight traffic and the need for a specific class to haul freight in these difficult circumstances became clear. Richard Maunsell took his 'N1' class 2-6-0 and used it as the basis for the 'W'. The cylinder diameter was increased by a ½ inch to aid acceleration and hill climbing and a high brake power ratio featured. The tanks, cabs and bogies – the latter brake-fitted - recovered from the ill-fated 'River' class were reused. The first five were built at Eastleigh and introduced from January 1932, No 1915 the last of this lot, in February. These had Ashford-style right-hand drive. Ironically perhaps, the second lot of ten came out of Ashford in the year from April 1935 and featured Eastleigh left-hand drive. 31915 was withdrawn in October 1963. *(A E Bennett)*

'P' class 0-6-0T No 31557 stands beside the BR-built signal box at Stewarts Lane in charge of a traffic now totally lost to the railway – milk - though this tanker is empty and awaiting transfer probably to Clapham Junction for the afternoon train to the West Country. The photo dates from September 1955 and though by this time traffic of milk in tanks predominated, the use of churns was still in being and did not finally cease until 1966. The Southern was the last of the 'Big Four' to introduce bulk conveyance of milk, four 4-wheeled trucks being turned out by Lancing Works in September 1931 to carry road tankers built by R A Dyson & Co of Liverpool for the Co-operative Wholesale Society. The underframes conformed to those of RCH 20-ton mineral wagons. Six more frames came from Lancing in October and November that year, to be fitted with United Dairies glass-lined tanks. Using a 4-wheeled chassis soon proved to be a mistake. At speed their riding was rough and unsteady, such movement creating the right conditions for the milk to turn into butter. On the basis that a tender full of water ran relatively steadily at speed behind an engine, the Board of Trade required six-wheeled frames to be fitted instead. The first of these left Lancing Works in October 1932, four of them to carry Express Dairies 3000-gallon tanks. It should be noted the tanks were the property of the dairy company but the railway retained ownership of the truck. Milk traffic along the ex-LSWR line to the south west ceased in 1967 following boundary changes and the singling of the route, west of Wilton. Thereafter, the trains went via Kensington Olympia to/from the Western line. No 31557 – originally SECR No 754 - was renumbered A557 in 1927. It lasted in service until September 1957. *(A E Bennett)*

A street view of Blackfriars station frontage in May 1972 shows a tired looking affair which still bears the soot from decades of smoky steam locos, as well as domestic and industrial chimneys. But it is just possible to make out the names of destinations reachable by rail that have been engraved into the stone mock colonnades - Bromley and Brussels are there, together with Sheerness and Lucerne. The gaudy paint scheme of the Escort van which is delivering the 'Evening Standard' could hardly be more of a contrast. Propped up against the wall are the Standard's headline boards which speak of a 'Rail Day of Decision'. This presaged a lengthy dispute between the rail unions and Ted Heath's Conservative government over pay. Tobacconists were a familiar sight on high streets across the land and John Gordon Miller established outlets at several railway stations; it is believed he was related to Yehudi Menuhin, no less and played a part in helping to launch the famous violinist's career in Britain after the war.
(Transport Treasury)

Faintly reminiscent of one of those old style Welsh dressers, this departure indicator board at London Victoria was built to last. And bearing in mind that it once read 'London Brighton and South Coast Railway' up above where now it only says 'Departures', then 'last' it clearly did. The 'Brighton Belle' Pullman service is showing on platform 4 so that would suggest this view dates from around 1970. Meanwhile, a blackboard to the right invites passengers to consider making a donation to the railway's Woking Homes charity for orphans and retirees whilst a trolley for a suitcase is propped up below; today, these two items have been combined of course and the trolley-case proves indispensable to many rail travellers. *(Transport Treasury)*

Left: At the height of summer, though the dull 12 August 1956 appears to give a lie to that, the engines at rest at Stewarts Lane shed include a Maunsell 'mogul', a SE&CR 'P' 0-6-0T, two LBSCR 'E2' class 0-6-0 tanks, a Great Western 'pannier' together with the tender of an anonymous 'Light Pacific'. The 'P', which appears to have an abundance of steam, is No 31325 of October 1910, a Wainwright design inspired by the Brighton 'Terriers'. The eight examples were introduced from February 1909, no 325 being the last. Originally intended for push/pull work they were soon found to be inadequate, later gravitating to light shunting and shed pilot duties. It is ironic perhaps that four of the eight survive in preservation. The 'E2s' came off Lawson Billinton's drawing board at Brighton in two batches. A commendably clean No 32103 was one of the original five introduced in 1913; 32105 came in a second batch two years later. In this 'lot' the top of the tanks was extended forward to increase capacity by 166 gallons: weight went up by 15 cwt. GWR Pannier tanks did not appear on local SR metals until 1959, and then at Nine Elms to relieve the ageing 'M7s' on empty stock work in and out of Waterloo. The presence of No 8757 here suggests it is visiting from the GWR goods depot at South Lambeth. *(A E Bennett)*

Right: 'Battle of Britain' class Pacific No 34090, photographed on 1 November 1951, has clearly not satisfied its last driver in some way, his report sending the engineers at Stewarts Lane to search for and right the problem. It was, perhaps, a good thing 'Health & Safety' rules were some way in the future for the stand looks rather unstable, with a stool on top of the platform and the fitter at full stretch. No 34090 was the last of twenty engines built at Brighton Works to Order no 3383, entering service in February 1949, fitted with a nine feet wide 'V'-fronted cab and hauling a 5,500-gallon tender (no 3340). First allocation was to Ramsgate shed. This 'lot' was also the first of the Light Pacifics not to be numbered under Bulleid's Continental system. 34090 bore the name *Sir Eustace Missenden, Southern Railway*, an acknowledgement of the crucial role played by the SR during the Second World War, though the original intention was to name it 601 Squadron, that name going to No 34071 instead. The naming ceremony took place at Waterloo on 15 February 1949; the engine was finished for this in SR malachite, wheels included, lost on reception of its lined dark green BR livery in March 1952. Even before completion of the first phase of the Kent Coast Electrification Scheme on 15 June 1959, No 34090 had left the Kent Coast for Nine Elms and the West of England main line. Reconstruction of the class on more conventional lines had begun with No 34005, *Barnstaple*, in June 1957, *Sir Eustace* undergoing the transformation at Eastleigh in August 1960. Having run just under 744,000 miles in service, withdrawal from that shed came at commissioning of the Bournemouth Line electrification in August 1967. *(Eric Sawford)*

On 11th March 1978 4EPB unit No 5330 on an Orpington-Charing Cross working has been routed along the Up Main but is here being crossed to the Up Slow on the approach to Hither Green whose 1933-built Loco shed, with its row of fuel tanks, may be seen beyond it. (If the headcode were '14' the train would travel via Lewisham and not the direct route this one takes via Parks Bridge Junction.) The first of these BR Standard units was turned out of Eastleigh in 1960 to supersede the now thirty-five year-old 4SUBs nos 4301-55. Numbered 5305-56 and sent to Eastern Section depots, the trailers in these were of unusual layout with five compartments and a five-bay saloon. Fourteen more units, Nos 5357-70, were built at Eastleigh in 1962 for the Western Section on frames made at Ashford, but they soon gravitated east, partly because their greater length over the SR version caused berthing problems at some of the country termini. (This batch could be identified by the shallower headcode panel.) Withdrawal began in the mid-1980s but many units were 'facelifted' and renumbered into the 56XX series, lasting to 1995. In 1972 the units became 423 under the TOPS system, amended in 1975 to 415/2; the facelifted sets were classified as 415/6. *(Bernard Mills)*

A bleak, cold February evening sees a 'Crompton', class '33', at London Bridge with a train for East Grinstead via Oxted. Diesel electric units of classes 205 and 207 formed the usual rolling stock for such traffic but these engines, with sets of seven or eight carriages, would appear in rush-hours or if a race meeting was being held at Lingfield. The train is standing in the tight space between the higher-level SER 'through' platforms and the massive yellow brick wall on the site of the London & Greenwich Railway terminus opened on 14 December 1836, the first in London. The Type Three 'Crompton', named after the manufacturers of the electrical equipment in them, has proved to be one of the most reliable and versatile engines that date from the 1960s, and many are still at work. *(Bernard Mills)*

Bulleid's 'double-deck' unit, No 4001 – very probably with 4002 attached - comes round the sharp curve into Cannon Street with yet another service from Gravesend via Bexleyheath. Though classed as 'double-deck' the two all-third sets were in reality better described as having 'stepped compartments', alternately high and low. Built experimentally to the absolute limit of the loading gauge their size effectively restricted them to the various Dartford routes. They came into service in 1949, each consisting, as usual with the then-current SR's units, of two motor coaches and two trailers offering a seating capacity of 552. Before the first year of service was out, it had become clear they were not the ideal solution to the overcrowding. The greater numbers of passengers boarding/leaving the train at stops as well as the inconvenience of internal stairs to negotiate, led to unacceptably lengthened station dwell-times. Ten-coach trains appeared the practical solution despite the expense not just of additional coaching stock but of lengthened platforms and, where necessary, repositioned signalling. Nonetheless, the two lasted in service until 1 October 1971. *(R C Riley)*

In late 1951 No 5001, classified as '4EPB', came off the production line at Eastleigh. Though the accommodation matched that of the later '4SUBs' and, like them, was mounted on salvaged underframes, it had buckeye couplings and electro-pneumatic braking and control gear powered by a motor generator beneath the floor. Brake and control jumper cables were mounted at cab height and on both sides, making it possible to couple and uncouple units from the platform. Access to the cab was now from the adjacent guards van and roller blind indicators replaced the long-used stencils. Nos 5002-15 appeared in 1952 and all were sent to work the Guildford 'New Line' via Cobham. Construction thereafter was virtually continuous until by the end of 1957, 213 units had come off Eastleigh's production line: they could be seen on all three Southern Sections. Five had, however, been written off or disbanded due to accident damage. The relatively rapid introduction was, perhaps, an indication that they were not compatible with the '4SUBs' which were withdrawn as these units took over the suburban area. Inevitably they all underwent a degree of refurbishment, being reclassified into sub-classes as a result. Pictured at Blackfriars on a bright, cold 18 January 1990 are classes 415/4 No 5460 and 415/1 No 5213. *(Bernard Mills)*

'4CAP' unit No 3301 stands in the bleak surroundings of Holborn Viaduct on a bright January morning in 1990. 2HAP units first left Eastleigh in 1957, this 'lot', numbered 5601-36, being known as '1951 Stock' because the motor coaches were the same as in the '4EPB' units of that year except for the saloon being divided into two. A second batch of similar stock, Nos 5651-84, followed in 1958/9. Forty-two more, numbered 6001-42, were turned out of Eastleigh in 1957/8 to supersede the '2HALs' on the Maidstone and Gillingham services, while another sixty-three, Nos 6043-6105, appeared in 1958 for the new electrification in Kent. The coaches in these were, however, of BR standard design and thus classified as '1957 Stock'. Change occurred in May 1982 when twenty-six '1951' and twenty-two '1957' sets were formed into '4CAP' units and based at Brighton for working along the Coast to Portsmouth. For this purpose they were permanently coupled in pairs with the motor coaches in the centre. Re-classification under TOPS was as 413/2 (1951) and 413/3 (1957), numbered respectively 3201-13 and 3301-11. All later moved back to the South Eastern area, based at Gillingham and Ramsgate from where withdrawal was completed in early-1995. *(Bernard Mills)*

Demand for off-peak travel into Cannon Street began to diminish from the later-1950s, to the point that twenty years later it was closed at weekends and during the evenings. Rumours spread in 1984 the station was to close, apparently confirmed by BR's 1984/5 Winter timetable showing all off-peak trains that would normally terminate there would do so at London Bridge instead. The rumours were denied, the point being made that experience showed the demand was not there. This was, however, a period of development in the City and the station site was seen as one begging for improvement as the 'property boom' continued. J W Barry's hotel frontage had been demolished and replaced with offices in the early 1960s, but this new period of development saw two office blocks built on a raft over the platforms. In the mid-2000s, the 1960s office frontage building was seen to have outlived its usefulness and was also demolished to be replaced by one of 'mixed use'. Full opening hours were still a little way in the future when these three class 465 'Networker' units, dating from 1991, were photographed on 27 November 2009.
(Andrew Royle)

The backdrop has barely changed since steam days but much else has as class 377 'Electrostar' unit, No 377155, in Southern livery, leads a companion up the bank from Victoria to Grosvenor Bridge and toward the south on 30 September 2011. Two more units occupy the long Up siding that parallels the line over the bridge, and a hint of a Gatwick Express may just be discerned passing to the rear of 377155. The yellow vehicles on the right and on which the photographer is travelling are forming a railway monitoring train. The first 377s emerged from Bombardier's Derby Works in 2001 to supersede the ageing - and not now meeting safety standards – class 4CEP and 4VEP stock working between the capital and the South Coast. Upgrades to the power supply to offset the greater demands of these units delayed entry until 2003. In all, 211 units had been brought into service before delivery ceased in 2014. The class is used for both suburban and mainline traffic and is divided into sub-classes, three of which, 377/2, 377/5 and 377/7 are dual voltage, such units working the local East Croydon-Willesden Junction service over the West London Line where the traction supply method changes north of Shepherds Bush. *(Andrew Royle)*

National Express began operating the Gatwick Express franchise under a fifteen-year contract in April 1996. However, as part of the award the company was required to provide new rolling stock in place of the class 73 locomotives that had been working trailers push/pull between Victoria and the airport for about twelve years. The class 460 units - 8GAT, numbered 460001-8 - were ordered by National Express and Porterbrook Leasing and came into service from Alstom from 2000. They proved unreliable however, and it required some extensive modifications before all were in service and the last of the ex-BR stock could be withdrawn in 2005. The lease passed to Southern when GA was merged with the South Central franchise in June 2008, Southern intending to use the 460s as part of an improved service south of Gatwick. The small number in the class required it to be supplemented by refurbished class 442 units. These gradually took over the Gatwick workings after renewal of the South Central contract in June 2009, permitting withdrawal of the class 460, completed by September 2012. In happier times unit 460002 passes through Selhurst on 22 April 2010. *(Andrew Royle)*

On 8 May 1979 a train very likely bound for Bognor Regis draws into East Croydon. The '4VEP' unit No 7911 that brings up the rear will not get to the coast but be detached at Gatwick Airport to let air travellers and their luggage leave it at relative leisure. Trains have served the new airport station since 1958, originally with a '2HAL' set detached/attached there to/from the half-hourly Victoria-Bognor Regis services. It was another twenty years before advocacy for a dedicated non-stop service began, though to start with the existing arrangements were updated by substituting class '423'/4VEP sets that had had changes to seating and luggage accommodation and were reclassified as class '427'/VEG as illustrated here. The service was not that successful, given the trains remained semi-fast via Redhill and could thus be overtaken by faster coastbound services calling at Gatwick. Real change came in 1984 with non-stop 'Gatwick Express' trains formed of Mk 2 coaches worked push/pull by class 73 electro-diesels. *(Bernard Mills)*

Inner- and outer-suburban stock side by side at Clapham Junction in the form of 4SUB unit No 4626 and 3-car diesel-electric set No 1316, on 25 January 1980. The latter was one of nineteen sets (Nos 1301-19) that first entered service in April 1962 for Oxted line workings though based and maintained at St Leonards West Marina depot alongside the Hastings diesels. No 4626 was toward the end of a long line of '4SUB' units going back to Bulleid's first six-a-side seating set No 4101 that appeared late in 1941. Like the older 1925 stock, this and its nine classmates had a canvas roof and domed front ends. Those fittings were lost in the next ten, Nos 4111-20, turned out in 1946. Instead an upright, slightly bowed front featured and the roof was now steel. Set No 4626 was one of forty-eight, Nos 4621-66, appearing in 1949 and built on salvaged frames recovered from withdrawn 3SUB sets. By this time the all-compartment internal layout of 4101 and its predecessors had given way to a saloon layout in three of the four coaches with one retaining compartments. These sets were quite long-lived, No 4626 surviving until October 1982. *(Bernard Mills)*

The 150th Anniversary of the opening of the London & Greenwich Railway was celebrated over the weekend 23/24 August 1986. Among the exhibits was a '4CEP' unit in 'Jaffa Cake' livery and a '4VEP' sporting the new Network SouthEast red, white and blue. 'N15' class 4-6-0, No 777, *Sir Lamiel*, was also in attendance and clearly attracting attention. A development of Urie's 'N15s' of the immediate post-WW1 years, 777 was one of thirty of the Maunsell version of the class, the 'Scotch Arthurs', turned out of the Hyde Park Works of North British between May and October 1925. The engine's preservation at the NRM comes from a 'normal' service run with the up 'Atlantic Coast Express' between Salisbury and Waterloo, but made in the theoretically impossible time of 72¾ minutes for the 83.8 miles, with a maximum speed of 90mph at Byfleet. Factors apparently in the engine's favour included a very strong Sou'-Westerly wind and an usually clear road throughout. Withdrawal and preservation occurred in October 1961 though the engine has since obtained a long and praiseworthy record of work on the mainline. *(Bernard Mills)*

The doyen of the twenty-four 2552hp Bo-Bo electric engines, Nos E5000-23, built by BR at Doncaster in 1958, and later classified '71', stands in Stewarts Lane yard. Like the Bulleid electrics that preceded them the bogies were not directly driven from the 3rd rail because of the possibility of 'gapping', in which none of the pick-up shoes were touching the rail, a real possibility in the complicated track layout outside termini. Instead, power was taken by a motor generator booster set which in turn fed power to the four 638hp English Electric traction motors through a flexible drive. A pantograph located in a cut-out section of the roof was also provided to take power from overhead lines erected in yards or locations, where laying a live rail was considered too dangerous for staff working at ground level coupling and uncoupling wagons, though these were later removed. In December 1962, in preparation for TOPS this engine was renumbered E5024, making the list E5001-24. Though noted as mixed-traffic engines their disadvantage was that much freight was moved at night, a time when engineering work might require power to be switched off. Similarly the class could not work inter-regional freight though this had now proved to be within the scope of the versatile class '73' electro-diesels and class '33' diesel electrics. Some class 71s were later modified through installation of a 600hp diesel engine to become class '74', but the rest, as class '71', were withdrawn by the end of 1977 and scrapped. No E5001 is preserved at the NRM. *(R C Riley)*

Bo-Bo electric No E5013 makes an effortless climb of Grosvenor bank with a train appearing to be predominantly of Maunsell stock. Introduced in 1958, the twenty-four engines in this class worked on the same booster system as applied by Bulleid to his Co-Co's first introduced in 1941. Weighing 77 tons, only three-quarters that of the Bulleids yet nominally 70% more powerful, these were very successful mixed-traffic engines with a reputation for swift acceleration, even with heavy loads. However, the later class '73' electro-diesels and class '33' diesel electrics both had the advantage of versatility whereas the '71' was restricted to 3rd rail routes. Eventually, there was not enough work for them that could not be covered by those classes and all twenty-four were gone by 1977, most in good working order. *(R C Riley)*

The great bulk of Hamptons Warehouse forms a backdrop to nicely turned-out Wainwright 'C' class 0-6-0, No 31293, standing at Stewarts Lane in company with a Maunsell 'N' 2-6-0. The 'C' was a development of the Stirling 'O' class by way of Wainwright's reconstructing many with the same style of boiler provided for the 'H' class 0-4-4T to become 'O1'. Wainwright himself had a greater concern for carriages than engines though he took a keen interest in boilers. It should not astonish then that with the Chatham's Chief Draughtsman, Robert Surtees, being appointed to the same office with the SECR after the 'fusion of 1899, the 'C' bore more of a semblance to the products of Longhedge than Ashford. No 293 entered service in June 1908 and spent the bulk of its existence at Bricklayers Arms shed, being withdrawn from there in May 1962. No 592 of February 1902 is preserved at the Bluebell Railway. *(Flint and Harbart)*

Two ex-SECR engines stand at Stewarts Lane with the huge Hamptons Warehouse providing the backdrop. Harry Wainwright had been appointed Locomotive Superintendent in 1898 upon creation of the two companies' Managing Committee, with the Chatham's Chief Draughtsman, Robert Surtees, moving to Ashford from the LCDR's works at Longhedge to continue in this position under Wainwright. From 1904 Wainwright ordered sixty-six 0-4-4T engines of the 'H' class, Not surprisingly, visually the design bore resemblance to the Chatham's 'R' and 'R1' classes, especially the later version ordered before formation of the Managing Committee, these having 'H' class boilers when outshopped by Sharp Stewart in 1899. The last of the class - or so it was thought – was completed in December 1909. Richard Maunsell, who succeeded Wainwright in 1913, somehow found the class was two short of the order and insisted it be made good. As a result, Nos 16 and 184 came into service in July and April 1915 respectively. No 263 went through a number sequence as ownership changed, twice by the SR, an 'A' - for Ashford – prefix changed to '1' from 1931, plus the BR 30000 post-1948. The engine survived until August 1960 when the 'H' Class Locomotive Trust purchased it on withdrawal for preservation, finally going to the Bluebell Railway in 1976. The 'C' class 0-6-0 ran to a total of 109 engines built over a period of eight years from June 1900 by Neilson Reid and Sharp Stewart as well as the SECR workshops at Ashford and Longhedge. Those from the last were probably erected from parts supplied by Ashford rather than built from scratch. No 31581 left Ashford in May 1903: surprisingly, Exmouth Junction featured among its allocated sheds. It was withdrawn from Stewarts Lane in March 1960 though one of its fellows survives, no 31592, a Longhedge-built engine from February 1902. After transfer to Service Stock as No DS239 in July 1963, it also went to the Bluebell post-withdrawal in May 1967, into the care of the Wainwright 'C' Class Preservation Society. *(R C Riley)*

Above: Steam Crane DS81 was photographed at Stewarts Lane in August 1966. It was one of a pair with 36-ton lifting capacity ordered in 1926 from the famous Ipswich firm of Ransomes & Rapier who built it under construction no C6553. It came into service the following year, numbered 81S by the Southern. (The other half of the duo was No 80S.) Allocated at first to Brighton shed, in time it moved along the coast to Fratton and then up the Direct Portsmouth route to Guildford, being renumbered DS81 by BR in the meantime. Stewarts Lane received it in 1963. Withdrawal came in 1986, sale to the Kent & East Sussex Railway at Rolvenden following in 1987. Its new owners subsequently downgraded it to 30-tons capacity. The twin, No DS80, was less fortunate, being scrapped in 1985. *(R C Riley)*

Right: 10 May 1959 finds BR Standard Class 5 4-6-0 No 73082 being prepared for traffic at Stewarts Lane shed. With the backdrop of Hamptons warehouse behind, it's a scene which simply oozes the atmosphere of the steam era, just as it is on the cusp of turning towards the modern era; new 'HA' class electric locos had already been recorded by the photographer here only a matter of weeks before. Footplate opinion of the Standard 5 was that it was a good engine but performed no better than Stanier's Black Five, even though it may have been marginally easier for the cleaning staff to dispose of. 73082 became one of the lucky ones to survive into the preservation era and now bears the name *Camelot* that it carried all too briefly during its Southern Region days. *(R C Riley)*

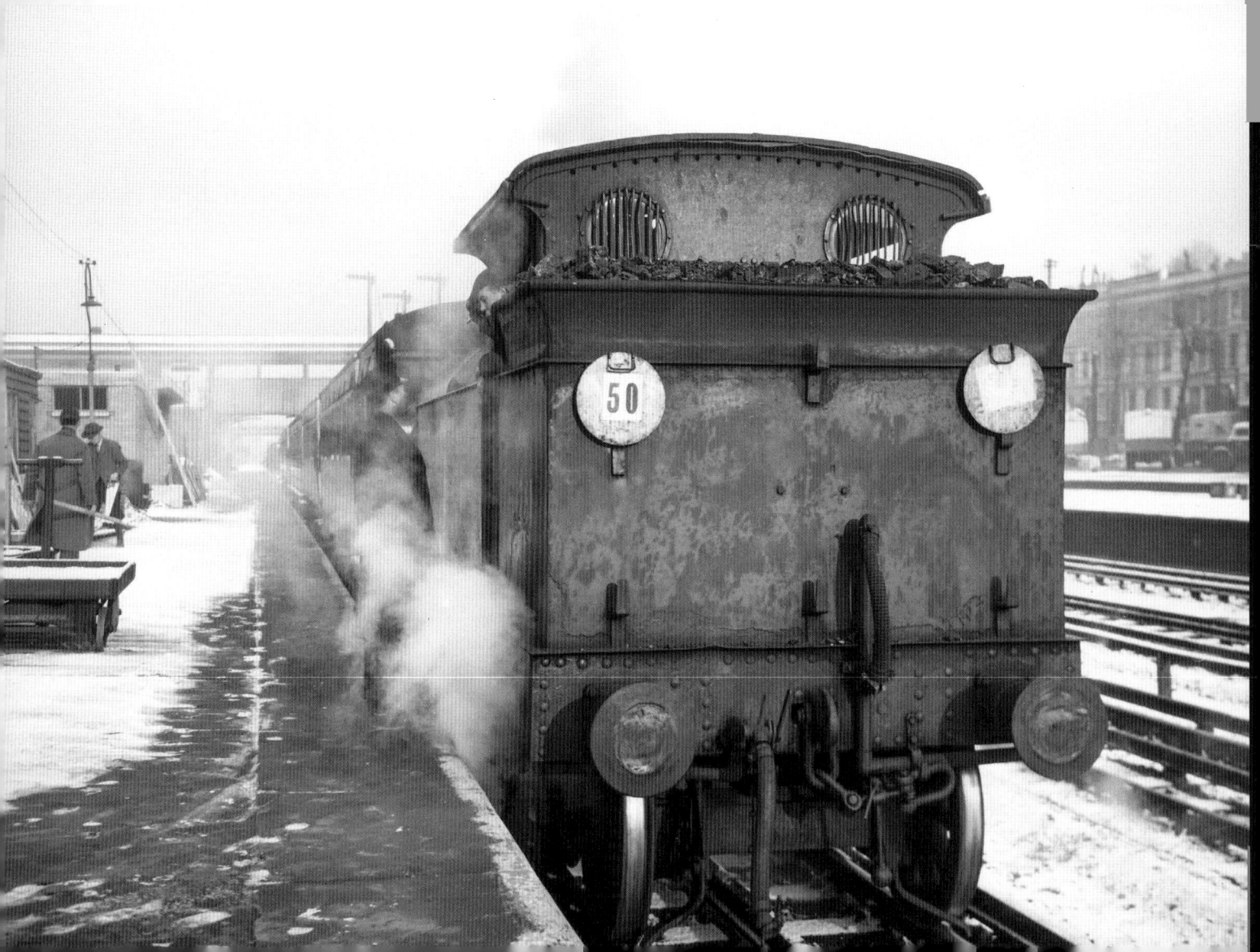

Left: For many years the West London Railway services were confined to goods traffic other than inter-regional passenger trains between northern cities and the South Coast passing through in high summer. The exception consisted of the limited shuttle between Olympia and Clapham Junction as represented here on 12 February 1955 by a train headed by Wainwright 'H' class 0-4-4T No 31261, about to leave with the 9.00am departure southwards. The service ran Mondays-Saturdays, though only the two early morning return trips worked consistently without change. Five other afternoon return trips are shown in the timetable but were restricted to particular days of the week, which resulted in only three being run on any one day. The engine, which displays the characteristic pagoda' roof fitted to most of the class, was built at Ashford in November 1905 and spent all but a few months of its existence at Stewarts Lane shed. Unlike many of its colleagues it never received fittings for motor working, meeting its end in October 1961. *(Leslie Freeman)*

Right Top: Standard '4MT' 2-6-4T no 80059 leads a train of Standard, Maunsell and Bulleid coaches into Clapham Junction on one of the regular 'eight minutes past the hour' trains for Tunbridge Wells West via Oxted. It was not until a complete recast of the Oxted line timetables in 1955 that regular 'clockface' off-peak departures of these services became common and concentrated at Victoria. Longer-distance trains were interspersed with these, making use of the line through Sheffield Park and Lewes to reach Brighton or Eastbourne, usually being the province of Maunsell 'moguls'. This engine, turned out of Brighton Works in March1953 as the first of Order no BR56271, was well travelled from Kentish Town when new to Chester and then Neasden. It probably came South with the many others exchanged for Brighton-built LMS Fairburn 2-6-4T returning to their spiritual home north of the Thames. No 80059 was one of the class posted to the S&DR in its last days, into the care of staff at Bath Green Park before withdrawal from there in July 1965. *(Transport Treasury)*

Right Bottom: The 'Morden Milk' empties come off the Sutton line at Wimbledon 'C' Junction headed by Bulleid' Q1' 0-6-0 No 33030, whose filthy exterior matches that of the stock. The train will be making for the Windsor side of Clapham Junction to form part of the regular daily 3.54pm service returning the empties to Exmouth Junction. From there they will be distributed to the various loading facilities in Devon and North Cornwall. The heavy milk traffic on the South Western main line ceased following boundary changes in 1967 which saw that part of the route west of Salisbury pass to Western Region and much of it subsequently being singled. Nowadays, milk no longer features in Working Timetables. *(Transport Treasury)*

Just another ex-LBSCR 'E4' class 0-6-2T, albeit rather far from home in the vast surroundings of the rival South Western's Waterloo. The companies were certainly uneasy bedfellows where their borders met – Guildford, Midhurst and Havant for example, though they shared locomotive accommodation at Fratton – but a number of these engines were moved to Nine Elms to assist the ailing 'M7' 0-4-4T engines that for years had been the mainstay of empty stock workings between the terminus and the carriage sidings at Clapham Junction. No fewer than fourteen were employed though not all at the same time, the process being phased, drafting examples in to supersede others at withdrawal. No 473 was among the earliest of the class to leave Brighton Works, in June 1898, going new to New Cross (with a 'Gate' suffix post-Grouping) for local suburban traffic. It returned to Brighton shed a year before the company became part of the Southern Railway and is pictured here on 10 September 1961, just over a year before withdrawal and the move into another phase of its life. The Bluebell Railway had an 'E4' among the list of engines it considered ought to be preserved, and selected No 473 from the few survivors at Nine Elms. By good fortune in its early Brighton days the engine had carried the name "Birch Grove", an estate quite close to Horsted Keynes station and by this time in the ownership of the publisher and Prime Minister Harold Macmillan. The engine arrived under its own steam at the Bluebell in October 1962. *(James Harrold)*

Stanier '5MT' class 4-6-0 No 45222 heads a Norwood Junction-Willesden freight through Kensington Olympia on 11th February 1961. The class had first appeared in 1934, following on from the same designer's class '6P5F' 2-6-0 of 1933, eventually numbering 842 examples, the last being introduced in 1951. They were a 'go anywhere, do anything' engine which often took them away from 'Home' territory as here (note the GWR lower quadrant signals!). By this time the West London Railway had had no regular passenger services for years other than 'through' trains between Regions at the height of the Summer Saturday timetable, but it did carry a very heavy freight traffic which included considerable quantities of milk. Currently, there is a regular hourly Southern service over the WLR between East Croydon and Watford Junction and approximately every fifteen minutes, an Overground service between Clapham Junction and Stratford. *(A E Bennett)*

BR Standard '4MT' 2-6-4T No 80068 runs round the stock of the 'Kenny Belle' at Olympia on 17 March 1961 in preparation for another shuttle trip back to Clapham Junction. Built at Brighton Works in August 1953, the last to Order BR5271, the engine was posted when new to Watford shed for working commuter services out of Euston and remained there for most of the 1950s. It would by this time have been at Stewarts Lane, though withdrawal in October 1966 was from Feltham. *(James Harrold)*

Left: One of the less famous Southern engines which served the railway for over forty years, Urie S15 4-6-0 No 30511 swings around the curve on the approach to Vauxhall with a train out of Waterloo. Post-war development of office blocks on one side of the tracks contrasts with the chimney pots on the other; the buildings on the right were largely demolished during the 1970s, leaving a surprising bit of open green space and even the modern developments at the rear of the train have since seen renewal. The photographer would have had a line permit to take this position and with regard to the electrified third rail and busy traffic, no doubt simply told to 'exercise care' and keep a sharp lookout! *(R C Riley)*

Right Top: For many years milk empties gathered at Clapham Junction from various points around South London to be returned to the West Country. This was a regular Bulleid Pacific turn, as likely to find a Merchant Navy engine at the head as was the 'Light Pacific' seen here. The main afternoon train by this time was for Exeter Central, leaving Clapham Junction at 3.54pm: here, No 34079, *141 Squadron* has charge. Beyond a certain loading, double heading would be necessary to lift the train up the steep climb from Point Pleasant Junction at Wandsworth, with its severe speed restriction on the turnout, to East Putney where this view was taken. *(Larry Fullwood)*

Right Bottom: One of the last of the 'Light Pacifics', No 34105, left Brighton Works in March 1950 bearing the name *Swanage*. It is seen here on 29 September 1961 passing Wimbledon 'C' box where the Southern-built line to Sutton turned away south. That line or something like it made an appearance before the First World War when the London, Brighton & South Coast, South Eastern and District Railways were all tinkering with ideas for the route, though it was the District that obtained powers. This line would have started from the company's terminal platforms at Wimbledon had war not intervened and the project dropped. 1922 saw a revival, this time with the Underground and a reluctant Southern in play, the former aiming to make it a further extension of the authorised line from Clapham Common. In the event, as part of a wider agreement between the parties, the SR took it on and the Underground terminated at Morden. The line had been opened throughout on 5 January 1930. *(A E Bennett)*

The Up 'Royal Wessex' passes through Vauxhall with rather leaky class '8P' Pacific No 35021 *New Zealand Line*, at its head on 11th September 1961. British Railways introduced the train on 3 May 1951 as one of several nationwide marking the Festival of Britain held that year. It can, however, be traced back forty years to 1911 and the running by the London & South Western Railway of the first non-stop train to Bournemouth, departing Waterloo at 4.10pm and scheduled to take a straight two hours for the 108 mile journey. It continued on to Weymouth with calls at Poole, Wareham and Dorchester, and carried a portion for Swanage detached at Wareham. World War One put an end to it but workings were resumed in Southern days with two trains each way daily. However, the growing importance of Southampton saw a stop there inserted in one of these in each direction. After cessation in WW2 reinstatement came about in 1945, but only one train each way and then not non-stop to Bournemouth. This became 'The Royal Wessex' with departure from Waterloo at 4.35pm and the first call now at Winchester; others followed at Southampton and Brockenhurst, arrival at Bournemouth being at 6.55pm. From 1964 the train was not provided with a set of dedicated stock and ceased to run following completion of the Bournemouth Line electrification in July 1967. (*James Harrold*)

A '4SUB' unit arrives at Waterloo at the end of a working from Hampton Court as 'Merchant Navy' class Pacific No 35026, *Lamport & Holt Line*, departs with the 'Bournemouth Belle' Pullman. The engine was the seventh of the last 'lot' of eleven built to Order No 3393 and turned out of Eastleigh in December 1948. Being post Nationalisation, *Lamport & Holt Line* never carried Bulleid's Continental SR-style number but the full BR 5-digit one from the beginning. It spent early years at Stewarts Lane, being displaced to Exmouth Junction in 1957, well before the Kent Coast Electrification would have demanded such a move. The photograph is undated but the state of the engine's 'finish' would imply this is not long before electrification brought the curtain down on steam out of Waterloo. By that time, 35026 would have been resident at Weymouth for some months, withdrawal from there coming in March 1967. *(Bernard Mills)*

This is the BR 'Riverside Special' run in conjunction with the Ramblers Association on 23 June 1957. It started from London Bridge and is seen here at Wimbledon, crossing to the Down Slow from the ex-LBSCR route via Peckham Rye. From here the train travelled to Weybridge whence it continued through Chertsey, where a stop was made to connect with a river steamer to take visitors to the Ramblers Association Rally at Runnymede. Windsor & Eton Riverside was reached by way of the now long-gone West Curve at Staines. The engine is 'N15X' class 4-6-0 No 32331 *Beattie*, one of seven LBSCR 4-6-4 'Baltic' tanks introduced by L B Billinton between 1914 and due to wartime restrictions in 1922. Electrification of the Brighton Line in 1932 and other coastal destinations later on meant there was no longer a place for these engines; they were rebuilt as 4-6-0 tender engines between 1934 and 1936. *(Ken Coursey)*

Viewed from the Gap Road bridge over the South Western main line east of Wimbledon, 'Standard' '5MT' 4-6-0 No 73162 is working Bournemouth Duty 393, actually noted as being for a 'Pacific'. This train will be the 15.35 to Bournemouth Central, due at the coast at 18.52, implying a number of calls en route. The South Western's power station for its electric services dated from 1915 though the left-hand of the two chimneys is a later one, the original having been destroyed by a WW2 bomb which damaged much equipment within the building. The District Line, over which the LSWR ran its first trains between Wimbledon and the junction with the District at East Putney, may be seen sweeping away northwards while a train of electric stock leaves the washer. The gradient faced by trains on the up slow to the flyover across the Through Lines is easily seen. This was completed in 1936 when track use was rearranged to remove conflicting movements on the approaches to Waterloo. *(Larry Fullwood)*

Left Top: From 1 January 1963, regional boundary changes put the ex-LSWR main line west of Wilton into the Western Region. There followed severe cuts to the service on the line and substitution of stopping trains for the expresses to Exeter, Plymouth and the West. A change of motive power was also imposed, the Bulleid Pacifics being superseded by WR diesel-hydraulic motive power in 1964. Here, the 1.0pm for Exeter is in the hands of 2200hp 'Warship' class (later class '42') B-B No D816 *Eclipse*, seen passing Wimbledon goods yard. The SR's trusty class '33' diesel-electric engines took over all Exeter services from October 1971 following BR's policy of standardisation upon diesel-electrics, though there had been increasingly frequent sightings of them in the interim. *(Transport Treasury)*

Left Bottom: On 18 January 1995 Pressed Steel-built class 117 three-car diesel-mechanical unit No L707 sets off from Clapham Junction's platform two for another of those irregularly-timed trips to Kensington Olympia. The thirty nine units, built between 1959 and 1961, were a development of British Railways class 116, built under licence and based on the Western Region, mainly for suburban work between Paddington and Reading. In the early-1980s some units were refurbished before transfer to the Birmingham area and later to beyond English borders for service in Scotland and Wales. In later years still the units remaining at Southall received NSE livery as here. The class was particularly long lasting, the final workings being in 2000, and as an indication perhaps of their durability no fewer than twelve sets continue in preservation. *(Bernard Mills)*

Right: Network SouthEast was just over a year old when '47' class Co-Co engine No 47307, viewed from the roof of Canterbury House, set out from Waterloo with the 1635 to Yeovil Junction. 1982 saw the Regional operations in place since Nationalisation in 1948 being split into sectors that covered specific types of traffic, the London and South East Sector taking responsibility for passenger services throughout South East England. It was anticipated the sector would be in a position to cover its costs, though day-to-day operation would still be in the hands of the Regions. L&SE became Network SouthEast in June 1986 with its new red, white and blue livery under the directorship of the dynamic Chris Green, who had successfully transformed ScotRail. Apart from the trains, much effort went into improving the presentation of the stations and the standards of service. Responsibilities changed again in 1991 when the Regions were disbanded and each Sector gained control of almost all the assets and operations within its boundaries. Things changed again from 1 April 1994 when NSE was disbanded in the lead up to Privatisation and its operations transferred to train operating units. *(Bernard Mills)*

Left: Bulleid had introduced two 1470hp Co-Co electric locomotives in 1941 and got a third, more powerful version in service by 1948. And it was the bogies provided to them that formed the basis, with an additional carrying axle, of those under the two diesel-electric 1750hp locomotives that began working express passenger duties over the Southern's main line to Exeter and Plymouth in 1951. The decision to build these engines dated to 1946, with the equipment to be supplied by English Electric; a third example rated at 2000hp followed on. Bulleid had no real enthusiasm for them, design work not being completed at Ashford before his retirement in 1949. Had he been really interested it is probable these would be labelled 'the pioneers' rather than the LMS twins Nos 10000 and 10001, which were in traffic by 1947. A wet and windy Vauxhall sees the first of the SR units, No 10201, gaining speed on the Down Main line with a train for Plymouth, a journey of roundly 5¼ hours duration. The problem for BR was that as diesel numbers and classes grew these engines were inevitably labelled as non-standard, resulting in withdrawal of all three in 1963. The bogie design was, however, perpetuated in almost exact form under the first ten production English Electric Type 4s. *(Ken Coursey)*

Right Top: With the withdrawal of steam, the service subsequently provided between Clapham Junction and Kensington Olympia featured several varieties of diesel power. Originally classified '2H' these diesel-electric units were built at Eastleigh from 1957. The first 'lot', numbered 1101-18 were for service over non-electrified lines in Hampshire bounded by Portsmouth Harbour, Salisbury, Alton and Southampton Terminus. All were reclassified 3H when demand saw an additional trailer added after 1958. Sets 1119-1122 appeared that year for services between Ashford and Hastings while four more sets, Nos 1123-6, came in 1959 to augment the Hampshire batch. Seven more three-coach sets, Nos 1127-33, for Reading-Salisbury working, were in service by 1962. Under the TOPS system all were reclassified as '205' and on 21 March 1991 one of the latter units, No 205031, in BR blue and grey livery, rested at the junction's platform 2 before another trip to Olympia. *(Bernard Mills)*

Right Bottom: On 23 September 1971 '4SUB' unit No 4131, seen here at Clapham Junction on a Waterloo-bound service from Hampton Court, has what appears to be a peculiar make-up. It was one of two 'put together' in 1969, formed of spare augmentation steel-bodied trailers from the unique seven-coach Oxted-line set no 900 – later 701TC - and bounded by what appears to be '2HAL' motor coaches. Withdrawal of Bulleid 4SUBs began in 1972 and all had gone by 1983, though both the units of this type would likely have gone quite soon in the process as they were not 'standard'. *(Bernard Mills)*

Two 4COR sets are neatly framed by Clapham Junction's 'A' box as they head for Waterloo. These units first appeared in 1937 when twenty-nine 4-car sets, numbered 3101-29, were provided for the electrification of the Portsmouth Direct line. The self-contained 6-coach Brighton line sets were not perpetuated: instead 'through' gangways were built into the sets' ends so that movement between sets became possible. This meant the headcode panel was off centre, giving the coach ends a rather one-eyed look which, with the inherent naval associations, gave them the nickname 'Nelsons'. Another batch was introduced for the Mid-Sussex electrification of 1938, of which No 3140 figures in the list of twenty-six units, Nos 3130-55. In later years the class could be seen on Waterloo-Reading diagrams and those along the south coast. TOPS titled them class '404' before their last Portsmouth Direct services ran on 31 July 1971. *(Bernard Mills)*

4VEP unit No 3405 passes Queens Road station Battersea on a Farnham-Waterloo working in March 1991. These units, reclassified 423 under the TOPS system, were built at BRs York and Derby Works and introduced in 1967 for semi-fast services on the newly-electrified Bournemouth line. The 3+2 seating proved ideal for commuter work but was less well received when used on express duties alongside the contemporary 4CIG sets with which they were compatible. The first twenty units, Nos 7701-20 were allocated to Bournemouth for the start of full electric working in July 1967 and, at that time, were the only sets with AWS equipment. In view of the secondary nature of the services they covered the livery was all over blue, though they were later liveried in blue and grey and later still in NSE colours as shown here. There were eventually 194 units in service throughout the Southern area, all of which were refurbished between 1988 and 1995. In 1978 twelve units, Nos 7788-99, were fitted with luggage racks in place of some seating, renumbered 7901-12, classified 427 and dedicated to Victoria-Gatwick Airport workings, usually forming part of semi-fast Bognor trains and being detached at the Airport station. In the Up direction the unit would form the front part of the train from Gatwick. Withdrawal was scheduled to be complete by 2005 though the last did not go till November that year. *(Bernard Mills)*

No 205101 is an unusual visitor to the sidings at Clapham Junction on 17 December 1994. Originally classified '2H' this was the first of eighteen diesel multiple units, Nos 1101-18, built in 1957 at Eastleigh for service over various non-electrified lines in Hampshire. They were, in effect, diesel versions of the 2HAP electric units under construction at the same time, formed of a motor brake 2nd and driving trailer composite seating thirteen 1st and 114 2nd class passengers. Unlike the emus the two coaches were joined by buckeye couplings and not the chains and central buffer common to electric stock. The units entered service in green livery, the first few working Salisbury-Southampton/Portsmouth Harbour trains on 16 September, all eighteen being in use when services between Alton and Andover and the coastal ports opened on 4 November. Strengthening to three coaches with the insertion of a Trailer 2nd began in 1959, No 1101 on 3 October though the general application was shortlived with several being reduced to 2H as a result of difficulty in timekeeping over the sharply graded route between Winchester and Alton. No 205001 was the first set to be turned out in NSE livery, in May 1988 and worked its last train on 9 December 2004 and was then sold to the East Kent Railway. *(Bernard Mills)*

Headed by No 50049, *Defiance*, the 0811 from Exeter is near the end of its journey to Waterloo as it passes Queens Road station on 30 July 1991. The engine entered service as D449 in December 1968 and became 50049 under the TOPS system in January 1973. Upon completion of the WCML electrification, *Defiance* was moved in May 1974 from the LMR Stoke division to the Western Region's Bristol Bath Road depot; another move to Plymouth Laira came nearly three years later. In 1987 it was finished in grey livery, renumbered 50149 and experimentally fitted with lower gear ratio class 37 bogies. It was set to work for Railfreight, hauling china clay and stone traffic from the far south west to London. Then it reverted to 50049 in 1989 and is pictured here only a month away from withdrawal in August 1991. *(Bernard Mills)*

The days of the GWR pannier tanks - let alone the venerable 'M7s' - chuntering around the carriage sidings at Clapham Junction are in the past and it is the class 33s that have what responsibility is left for movement of ECS to and from Waterloo. No 6542 may be preparing for such a function under the watchful eyes of those in charge of Clapham Junction 'A' box, now in proper order after its partial collapse in May 1965. From 4 October 1971 the 33s took over haulage of Waterloo-Exeter St David's trains upon withdrawal of the 'Warship' class that had had them since the line west of Wilton became Western Region responsibility. As this photograph is dated September 1971 No 6542 may, in a month's time, find itself taking on the 170+ miles to Devon's County Town. *(Bernard Mills)*

Two trains head for Waterloo in the shadow of the ramp carrying Eurostar traffic to and from the ex-LCDR main line and the Channel Tunnel. The 5-car class 442 on the up main was one of twenty-four 5-car sets based on the Mk3 23m long bodyshell, fitted with air conditioning and plug doors. The contract was awarded to BREL and though the units were new, much of the traction equipment came from 4REP motor coaches introduced at the Bournemouth Electrification of 1967, as they were withdrawn. Service began in mid-1988 with running through to Weymouth in winter 1988, though it was not until 4 September 1989, when all sets were available after snagging problems had been solved, that the full service came into operation. The 442s continued mainly to/from Weymouth, though they began working to Portsmouth in April 1992. Internal refurbishment began in mid-1999 and completed in March 2001. Franchise renewal and financial changes saw the first withdrawals from October 2006 to the conclusion of the lease in February 2007. *(Bernard Mills)*

A foreigner in the Camp! A GW HST set passes the Covent Garden Market at Nine Elms working the 10.46 Waterloo –Cardiff on 4 May 1995. It is probable this was an example of a connection being made at the London terminus with a Eurostar train from Paris. Its position here indicates it will take the West London Line at West London Extension Junction and leave it at North Pole Junction to join the GWR main line at West London Junction. Such close and easy Continental connections were naturally severed when the HS1 line to St Pancras opened in 2007. *(Bernard Mills)*

The headcode borne by 'S15' class 4-6-0 No 30499 reads 'Waterloo or Nine Elms and Southampton Terminus direct (not boat trains)', However, the train, passing Wimbledon yard, is more suggestive of one due to terminate at Basingstoke, calling at Woking and all stations thereafter. Other than the two vans at the rear, a 'loose' coach has been added to the Bulleid '3-coach 'M' set No 799, a type much allocated to Oxted line traffic but delivered new to the Western Section in 1959, barely two years before this photograph was taken on 29 July 1961. The 3-set consists of two Corridor Brake 3rds Nos 4217/8, and Corridor Composite No 5779. (The 3rd brakes went to the Western Region in 1967, the Compo being withdrawn that year). The engine was to the design of the LSWR's engineer, R W Urie and among the first of its class to leave Eastleigh Works, in May 1920. Feltham shed claimed ownership from start in May 1920, to finish in January 1964. *(A E Bennett)*

Contrasts at Feltham shed on 9 October 1954 with Drummond 'M7' class 0-4-4T fronting a Urie 'S15' 4-6-0. The tank engine was eighth of ten produced at Nine Elms to order E9, appearing between January and May 1899, this example in March. It was allocated to Feltham by the end of 1950 together with its class companions Nos 30249 and 30254, and remained there until withdrawal in May 1961. The class, which numbered 105 in all – the last not being produced until December 1911 and then from the new Eastleigh Works – could be seen all over the South Western system and even further afield after Grouping in 1923 when ex-LBSCR sheds such as Brighton and Horsham housed them, particularly for push/pull working. The 'S15' dates from 1920, No 514, the penultimate member of ten, leaving Eastleigh in March 1921. Richard Maunsell made modest changes to the class in a 'lot' of fifteen outshopped from Eastleigh between March 1927 and January 1928, five of which were fitted with 6-wheeled tenders for service on the Central Section. Another ten came into use in 1936 when Maunsell's plans for a 3-cylinder 2-6-2 mixed-traffic engine were turned down by the Civil Engineer. No 30514 was always a Feltham engine until withdrawal in July 1963; its classmate No 30506 is preserved at the Mid-Hants Railway. *(Leslie Freeman)*

The Eurostar terminal and its approach viaduct are seen under construction on the site of the old 'Windsor' station at Waterloo in July 1992. No doubt to the annoyance of local motorists, a goodly section of West Road is boarded off to provide a safer environment for the workers and permit materials to be delivered. Elsewhere, a number of platform lines are occupied and a 4CIG (class 421 under TOPS) arrives, possibly from Portsmouth. Despite the apparent progress that has been made, it will be two years and more before the first Eurostar service comes down the ramp at Nine Elms and runs beneath the Architects Grimshaw's superb asymmetric roof. *(Bernard Mills)*

What is the Penzance sleeper train behind 47815 doing at Waterloo? This photo dates from July 1996, a time when the Sleeper used Waterloo as its London terminus to provide a direct connection with the Eurostar services before their transfer to St Pancras. This train had a long history beginning in 1877 with a – naturally – Broad gauge service to Plymouth. Another train, to Penzance, was soon added, eventually known as the 'Night Riviera'. The two continued to run as separate trains two hours apart until after Nationalisation when sleepers were limited to the Penzance train, though sleeping cars were detached from it at Plymouth until 2006. New Mk3 air-conditioned sleeping cars were introduced in July 1983 and the service became part of the GWR franchise in February 1996. Stock was refurbished in 2017/8 to include wheelchair space and an accessible berth and toilet. Presently, the train picks up at Reading and then runs non-stop to Taunton, which makes a number of routes available between the two in the event of engineering works or a blockage. *(Bernard Mills)*

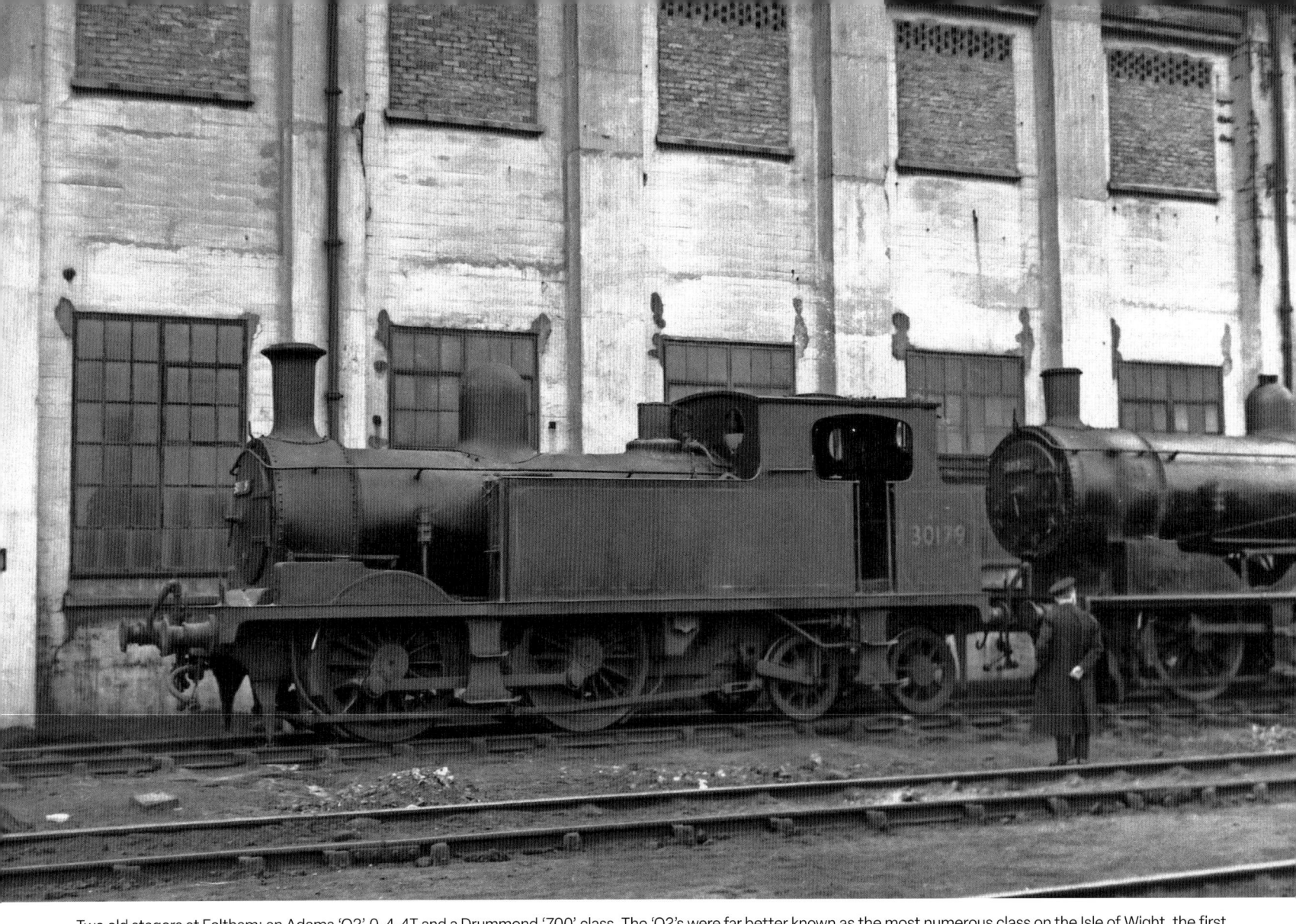

Two old stagers at Feltham; an Adams 'O2' 0-4-4T and a Drummond '700' class. The 'O2's were far better known as the most numerous class on the Isle of Wight, the first of them, Nos 206 and 211, crossing the Solent in May 1923. Twenty-three of the engines made the trip, the last two, Nos 181 and 198, as late as 1949. (Island engines were placed in a separate number list.) All were fitted with airbrake gear- relevant on the Island railways - before transfer and, from 1932, with the big MacLeod bunker to double capacity and so provide day long availability at the height of the season when some duties demanded the engines run more than 200 miles in a day. Those that remained behind seemed 'dainty' by comparison though until the 'M7' came to monopolise the work, a group of 'O2's were regular workers on heavy empty stock trains between Waterloo and Clapham Junction. No 30179 was the third member of the class, leaving Nine Elms Works in March 1890. It is seen here at Feltham on 11th October 1958, just over a year before withdrawal. *(A E Bennett)*

Steam is nearing its end at Waterloo, epitomised by the Western Region maroon liveried 'Warship' diesel-hydraulic D869 *Zest* in the sidings. A distinctly grubby BR Standard class 2-6-4T No 80144 waits to take another train of empty stock to the sidings at Clapham Junction. The engine was one of the last twenty built at Brighton to Order No BR7739 and began work at Neasden in May 1956. Steam was eliminated from Marylebone in 1966 but years before then No 80144 had crossed London to Nine Elms. The photograph was likely taken very shortly before withdrawal in May 1966 after a life of not quite a decade. *(Bernard Mills)*

'G16' 4-8-0T No 30495 stands beside the breakdown crane in the shadow of Feltham shed. Construction of the marshalling yard at Feltham and the shed to serve it arose from the growing pressure on the limited space at Nine Elms goods yard and the rising costs imposed by the London & North Western Railway for sorting and marshalling wagons on the South Western's account at Willesden. The greenfield site on the south side of the Windsor line west of Twickenham was purchased in two lots, a total of seventy-nine acres, the yard opening between December 1917 and 1921 with hump shunting mainly employed. Four examples of the 'G16' were built to the designs of Robert Urie to propel trains of wagons over the humps to run by gravity into the sorting sidings. No 30495 came into service in August 1921 and was withdrawn in December 1962, having spent its entire career at Feltham other than a few months at Strawberry Hill before Feltham shed was ready. *(R C Riley)*

A typical view of Nine Elms yard, where Bulleid Pacific No 34009 receives attention from a member of the shed staff on 8 March 1964. The engine came into service from Brighton Works in September 1945 as one of twenty built to Order No 2421. It first carried the number 21C109 and was named *Lyme Regis* on allocation to Exmouth Junction shed: renumbering occurred in March 1949. It came into traffic in 'rebuilt' condition in January 1961, having already returned to London and Nine Elms where it remained until withdrawal in October 1966. It was one of several 'Light Pacifics' that went to J Buttigieg of Newport and had been broken up by December 1968. *(George W Smith)*

Below: One of the more overlooked locomotive classes was the 'K' class 2-6-0 designed by Billinton for the LB&SCR in 1913. 32353 was the final one of the 17 built and is captured passing North Pole Junction with a northbound summer Saturday extra on 8 August 1959; still proving its usefulness in traffic right up to the point of withdrawal in late 1962. The train was probably not travelling at great speed, judging by the fact that the signalman has already returned the relevant signal to danger. Also apparent is the variety of rolling stock within the train, something which was often an indication of a summer extra. Back then, BR would keep all manner of vehicles in reserve to meet peaks in demand; it was a profligate use of resources that would be severely pruned under Beeching and continue to decline in the years that followed. *(R C Riley)*

Right: Not quite from such a high vantage point as Southwark Towers, though equally interesting is this scene overlooking the signal box at Waterloo. This featured prominently in the British Transport Films production 'Terminus', directed by John Schlesinger. 'West Country' Pacific No 34029 *Lundy* pulls out from the terminus on a fine but slightly hazy day. The lack of any train in platforms 1 through 5 suggests that this is a quieter time of day with most commuter trains stabled in sidings elsewhere, waiting for the evening peak. Today's railway offers a more regular 'clockface' style of service and likely sees few platforms completely empty for very long. Luckily for the railways (BR and post-privatisation), the LSWR had invested in a total rebuild of the terminus that was completed in 1922, providing the much needed capacity that 21 platforms could offer for the growth in passenger numbers that would come later on. Having also expanded into the former Eurostar station alongside, further enhancements to the site are currently being proposed. *(R C Riley)*

Urie's 'H15' class 4-6-0 mixed traffic engines had first appeared in January 1914 and ten were in service by the end of that year. They followed Drummond tradition in being of very solid 'battleship' construction but in all else differed. Gone were the deep driving wheel splashers, wing plates, firebox water tubes, patent – and not especially effective! – steam dryers – and in came generous bearing surfaces, big cylinders with decently proportioned front ends, superheaters and free steaming boilers. Urie had also rebuilt (in the December) Drummond's monstrous 'E14' class 4-6-0 to such an extent it was more or less new, being assimilated into the 'H15'. Ten more 'H15s' appeared under Southern auspices in 1924 and the five Drummond predecessors to the 'E14', the 'F13s', also found themselves in the last two months of 1924 carved up into more useful specimens and included in the 'H15'. No 30477, seen here 'on shed' at Nine Elms on 12 July 1954 features the great arched roof to the cab very typical of Urie's designs, which restricted such engines to the Western Section. At this time the engine was allocated to Eastleigh from where it was withdrawn in June 1959. *(Eric Sawford)*

The unmistakeable 'Pullman' livery of the 'Bournemouth Belle' contrasts very favourably with the dull black of class '4MT' 2-6-4T No 80145, whose lining is almost lost in the dirt. It will be another year before steam and the 'Belle' are eliminated together. The LSWR first dipped a toe into the Pullman pool in the mid-1880s with one coach in the morning service to Exeter. It failed to attract interest but another inserted in 1890 in the 12.30pm, to Bournemouth proved much more successful until, in 1905, a car featured in four trains daily. However, the introduction of corridor vehicles and restaurant cars saw a decline in demand until Pullman service on the South Western ceased in 1911. The Southern introduced the 'Bournemouth Belle' in July 1931, at first on Summer Sundays only though by 1936 its popularity had made it a year round daily working. In summer 1953, as the service recovered from the war years, it departed Waterloo at 12.30pm with a running time of two hours and ten minutes to Bournemouth Central, including a three-minute stop at Southampton Central. In the last few months, before withdrawal on 9 July 1967, haulage was by Brush Type 4 diesel. *(Bernard Mills)*

Six members of the GWR '5700' class 0-6-0PT were moved to Nine Elms from South Wales in 1959 to take on some of the ECS duties covered by the aging 'M7' class 0-4-4T. They were not wholly unknown in this part of London for engines of the class covered much of the transfer and shunting work at the Western's extensive goods depot at South Lambeth, tucked into the space between the Battersea Dogs' Home and the power station. Here, one of the six, sporting a disc with Nine Elms Duty 60 on it, brings a train past Queens Road – now Queenstown Road – station Battersea, with Bulleid 5-Set No 838 leading. This was introduced with eleven others in June 1950 in time for the Summer season traffic. It consisted of two Brake Thirds, two Composites and a Third. Earlier sets came with three coaches strengthened to five by 'loose' Thirds as required, but Nos 838-49 were always of five coaches. The set lasted in this formation until 1965 when it was, with most of its type, disbanded. The Panniers were never particularly popular, presumably based on the long lasting rivalry between the two companies for West of England traffic and an unfamiliar driving position. They were allocated away by the end of 1963 after Nine Elms acquired a number of standard class '3MT' 2-6-2T engines in their place. *(Transport Treasury)*

Left: An RCTS Railtour around the Southern's suburban branches in December 1962 brought the last remaining Beattie well tanks from their remote Cornish home to London. Nos 30585 and 30587 are pictured back-to-back receiving refreshment at Waterloo. Remarkably these two, together with 30586, remained after all the other members of the eighty-five strong class had been withdrawn by 1898 because no other suitable motive power could be found for working the Wenford branch. More remarkably still these two survive in preservation, No 30585 at the Buckinghamshire Railway Centre of the Quainton Railway Society while No 30587 is in the ownership of the National Railway Museum but based at its long-term home under the care of the Bodmin & Wenford Railway. *(R C Riley)*

Below: A West of England express formed of Maunsell stock in green but led by a crimson and cream 4-set of BR Mk 1 coaches, passes Vauxhall on 20 June 1959 under the charge of the last 'West Country' Pacific No 34108 *Wincanton*, built at Brighton in April 1950 as part of Order no 3486. (Under this same order ten of the Light Pacifics were built at Eastleigh, the only members of the type not to be turned out by Brighton Works.) Along with classmate No 34104 *Wincanton* was the last of the Light Pacifics to be rebuilt, in May 1961. Withdrawal came from Salisbury shed in June 1967. *(R C Riley)*

This is the view over the Up Reception Sidings of Feltham Marshalling yard from the Up Hump box with the Windsor Lines in the background. Typical old and (relatively) new motive power on 11th October 1958 is represented by a Urie 'S15' class 4-6-0, Bulleid 'Q1' class 0-6-0 and the inevitable English Electric diesel shunter. Plans for the yard were started in 1910 with land purchased in two lots in 1911 and 1915 to a total of seventy nine acres. Among other things the River Crane which crossed the area from Hounslow Heath had to be diverted and covered. The first sidings were opened in December 1917 but it was not until 2 October 1921 the yards were fully commissioned. However, the engine shed had not been completed at that time and until it was, in 1924, locos were handled at Strawberry Hill. The loss of freight, particularly wagonload traffic, from the late 1950s saw the yard become redundant up to its closure at the beginning of 1969. After a long fallow period it now houses Royal Mail's centre for handling its business over a wide area of South and South West London, and in late 2020 South Western Railway completed a depot for Class 701 units working the Reading services. (A E Bennett)

'Merchant Navy' 4-6-2 No 35029, *Ellerman Lines*, and 4SUB unit No 4754 race for Clapham Junction on 10 September 1962. The Pacific, the penultimate in the class and built as part of Order HO3393, left Eastleigh Works in February 1949 for Dover shed and its share of Boat Train traffic. Spring 1957 had brought it to Nine Elms though its final destination was on the coast at Weymouth whence withdrawal occurred in September 1966. Rescued from Woodhams yard at Barry it is now in the ownership of the NRM and suffers the indignity of being displayed sectionalised for educational purposes. The 4SUB was the eighty-seventh and last of the final lot which appeared in 1950/1. A few were mounted on recovered 3SUB frames but each set included a salvaged Bulleid augmentation compartment trailer, being otherwise of saloon type. *(R C Riley)*

Left Top: Waterloo & City Railway car No S60 has been lifted from the subterranean depths at Waterloo and taken to the Works at Selhurst for refurbishment where it is seen on 23 February 1975. Designed by the Southern in a rather Art Deco style, the twelve motor coaches and sixteen trailers were built by English Electric and introduced in 1940. Trains were formed of two motor coaches sandwiching three trailers but unlike other Southern stock these were not fixed and numbered formations. Outside rush hours, when demand was normally very slack, it was usual to find a motor coach running as a single unit. BR classified these vehicles class 487 and though resembling London Underground stock their shorter length was dictated by that of the hoist at Waterloo. From introduction until the late-1960's livery was green with aluminium ends and doors but BR blue followed, as seen here, and then Network SouthEast livery which saw ends and doors painted. All twenty-eight vehicles were retired in 1994 following introduction of Central Line 1993 stock on 'The Drain'. *(Graham Taylor)*

Left Bottom: Displaying its proper '62' headcode, No 33114 is near the end of its journey with the 12.13pm service from Salisbury as it passes Queens Road on 21 March 1991. This Sulzer-engined class was a development of the class 26, lacking the train heating boiler but with a more powerful traction unit to match the Southern Regions particular needs. Introduced from 30 January 1960, eighty-six units, later classified 33/0 and 33/1 under the TOPS system, were supplied in two batches by BRCW, originally numbered D6500-85. Later, twelve more, with narrower bodies, classed 33/2 and numbered D6586-97, were built for the width-restricted Hastings line and nicknamed 'Slim Jims'. *(Bernard Mills)*

Right: Former type '3', now class '33' No 33105, stands in a wet Waterloo with a train for Salisbury. The class was the first in the UK to be fitted with electric train heating apparatus and the first main line one to be provided with duplicated controls on each side of the cab. From introduction in January 1960 its existing versatility grew with demand, being capable of working with the Southern's electric stock and the class '73' electro diesels as well later being equipped in 1967 for push/pull working on the Bournemouth – Weymouth line. As an indication of both versatility and reliability no fewer than twenty-nine units, roundly 30% of the class, remain in preservation. *(Bernard Mills)*

Class 47 47715 *Haymarket* finds itself 'piggy in the middle' as it brings empty stock into Waterloo for the 17.05 to Exeter St David's on 21 July 1992. It is flanked by a '4VEP' – once described by a railwayman as 'a 4SUB with corridors' – introduced in 1967 for the Bournemouth Line electrification of that year. Beyond is a class 455 set built by British Rail Engineering Ltd in York and first introduced in 1982 for services over the suburban lines of the former London & South Western Railway. 137 sets were constructed though one coach in each was taken from a class 508 set before transfer to Merseyside for whose services they were originally intended. In true Southern tradition the 455s survived for many years, the final withdrawals being in 2022. The 4VEP was also a product of BREL York though the Motor Brake Seconds and Open Seconds in the first twenty were built at Derby Works. Altogether 194 sets were built to various orders up to 1974 and became class 423 when TOPS was introduced. Twelve units were modified in 1978 to provide extra luggage space and reclassified as 4VEG/427 for the Gatwick Airport. Service. In 1984, when the Gatwick Express was introduced they were returned to normal state and regained their original numbers. Despite their general lack of popularity, falling into the gap between suburban and main line, the last run was not until 7 October 2005. *(Bernard Mills)*